THE MOLE IN MEDICINE AND BIOLOGY

THE AUTHORS

J. A. OWEN
BSc, MD, ChB(Edin), PhD(Edin), MCPA, MC Path.
Director of Biochemistry, Alfred Hospital, Melbourne

R. G. EDWARDS
MSc(Lond), MBBS(Adel), MCPA, MC Path.
Head of Division of Clinical Biochemistry, Institute of Medical and Veterinary Science, Adelaide

B. A. W. COLLER
BSc, PhD(Melb).
Senior Lecturer, Monash University, Melbourne

THE MOLE IN MEDICINE AND BIOLOGY

By

J. A. OWEN

R. G. EDWARDS

B. A. W. COLLER

E. & S. LIVINGSTONE

EDINBURGH AND LONDON

1970

SBN 443 00650 4

Printed in Great Britain by
Spottiswoode, Ballantyne and Co Ltd
London and Colchester

PREFACE

The mole is a fundamental unit expressing the amount of a chemical substance. Students of chemistry at School and University have been taught to think in terms of moles for some years past but, in medicine and biology, other units such as the gram or the equivalent are still in common use.

The purpose of this booklet is to provide an account of the mole concept, to show how the mole is related to other units and to illustrate how it can be used to advantage as a basic unit in everyday medicine. It has been written principally for those in medicine whose formal training included only brief contact with the mole concept and who now wish to gain working knowledge of the mole as a unit. It may also serve as supplementary reading to students meeting the mole for the first time.

J.A.O.
R.G.E.
B.A.W.C.

1969

TABLE OF CONTENTS

1. INTRODUCTION

A variety of different units are currently used in describing the chemistry of living things. Mammalian body fluids, for example, are characterised in terms of inorganic ions computed in milliequivalents, organic substances such as protein, or glucose computed in grams or milligrams and enzymes and hormones computed in variously defined units.

Growing up with these different units and using them daily, physiologists and clinicians seldom find occasion to question their basis. In the more fundamental sciences, however, the shortcomings of these units are now well established. For the past few years, students of chemistry at school and university have been taught to think of quantitative chemistry in terms of moles, rather than in equivalents or grams, and it is inevitable that usage in medicine and biology will come into line with usage in these more fundamental disciplines sooner or later. Its acceptance as one of the International System of Units (S. I. Units) has been recommended by the Commission on Symbols, Terminology and Units of the Union of Pure and Applied Chemistry (IUPAC, 1967) and by the Commission on Clinical Chemistry (Dybkaer and Jorgensen, 1967) and many papers have already appeared in the world medical literature expressing biochemical data in terms of moles.

The aim of this manual is to give an account of the mole concept, to show how the mole is related to other units and to illustrate how it is used as a basic unit in everyday medicine. We have done this primarily for those in the biological sciences whose contact with the mole concept has so far been limited and who wish to gain working knowledge of the mole as a unit. The principles of the mole notation are detailed in the chapters which follow but we feel it would be helpful to the reader at this stage to summarise the benefits to be derived from using the mole as a unit.

The use of the mole as the fundamental chemical unit in physiology and medicine brings terminology in these sciences more into line with that in chemistry and physics. To some, this alone is sufficient reason for accepting the change, for the flow of knowledge from one branch of science to another is undoubtedly hindered if there are language barriers or communication problems. There are, however, direct advantages in using moles in practical everyday medicine.

To start with, expressing the concentration of chemical components of body fluids in terms of moles goes a long way towards providing a uniform system of physiological concentration units (Tables I & II). Although enzyme concentrations cannot usually be expressed in terms of moles at present, the expression of plasma enzyme activities in International Units (μmoles of substrate converted per unit time per litre of plasma) at least brings enzyme concentrations into line with those of simple substances.

Secondly, use of the mole scale avoids the ambiguities which arise in the use of the equivalent or of its subunit, the milliequivalent (p. 13). For instance, it resolves the problem of concentration units for polyvalent ions, such as phosphate, whose degree of protonation and, consequently, effective equivalent weight depend on the pH.

Another advantage of the mole scale over the equivalent scale is that it allows ready appreciation of the osmolal significance of plasma components and ready calculation of the osmolality of a fluid of known composition (p. 34). This is important in situations where direct measurement of osmol-

TABLE I: THE COMPOSITION OF NORMAL HUMAN PLASMA

The data comprises a consensus of the reports of various authorities.
The Table is not intended as a working list of normal ranges, but only as an illustration of how the concentrations of various components in plasma can be tested on a uniform scale (see text).

Substance	*Molecular or Ionic weight*	*Concentration* (mmol./*l.*)
Albumin	68,000	0.5–0.6
Ammonia	17	20–40
Ascorbate	176	10–30
Bicarbonate	61	20–26
Bilirubin	585	2–20
Calcium	40	2.3–2.8
Chloride	35.5	95–100
Cholesterol	387	5–9
Copper	63.5	15–20
Cortisol	362	0.2–0.8
Creatinine	113	6–9
Fatty acid (as stearate)	284	6–15
Glucose	180	4–6
Iron	56	males 12–36 females 12–45
Magnesium	24	0.7–1.0
Phosphate, inorganic (as P)	31	0.8–1.4
Potassium	39	3.5–5.0
Sodium	23	135–150
Urea	60	2.5–8.0
Uric acid	168	males 0.3–0.5 females 0.2–0.4

ality is not possible. The significance of osmolal changes in the plasma in relation to cerebral function is now well established.

In therapeutics, where dosage in moles, or in a suitable subunit, avoids confusion when pharmacologically active substances may be supplied in several different forms. For

TABLE II: NORMAL EXCRETION RATES OF URINARY COMPONENTS IN TERMS OF MOLES

Some excretion rates are influenced by dietary variations.

Substance	*Molecular weight*	*Urinary excretion rate* (mmol./24 hr)
Calcium	40	2.5-7.5
Copper	63.5	less than 0.8
Creatinine	113	10-16
5-Hydroxyindole-acetic acid	191	0.02-0.05
17-Hydroxycorticoids*	362	15-60
17-Ketosteroids†	288	Males 30-60 Females 20-45
Lead	207	0.12-0.5
Magnesium	24	5-20
Methoxyhydroxy mandelic acid	198	0.02-0.04
Phosphate, inorganic (as P)	31	10-30
Potassium	39	50-100
Sodium	23	100-150
Urea	60	400-500
Uric acid	168	2-5

* As hydrocortisone.
† As dehydro-epi-androsterone.

example, a dose of histamine (mol. wt 111) contains almost twice as much histamine as the same weight of histamine dihydrochloride (mol. wt 184) and three times as much as the same weight of histamine diphosphate (mol. wt 307). Prescribing a dose of histamine in millimoles (mmol.) would avoid this confusion for such a dose would specify a given amount of histamine regardless of the histamine salt used. Similarly 100 mmol. of calcium could be supplied as a solution of 15.8 g. of acetate, 21.9 g. of hydrated chloride, 30.8 g. of lactate or 44.8 g. of gluconate.

The transition from units currently in use to moles or

millimoles will undoubtedly give rise to confusion but this will be transient; moreover, as in the case of change to a decimal currency system, the longer the change is postponed, the more expensive it becomes. We cannot claim that introduction of the mole into everyday clinical chemistry will lead to immediate advances in patient care. It is certain, however, that use of the mole concept in medicine will, in the course of time, lead to a better understanding of the chemical processes which make up life and, as a consequence, to a better understanding of the nature of disease and needs of patients.

2. THE MOLE CONCEPT

2.1. The mole as a unit

The idea that matter is made up of discrete particles has its roots in history. Much of the evidence which led to the development of modern atomic theory came from quantitative studies–from measurement of the weights of substances reacting together and of the weights of products formed. Nowadays, chemistry is almost entirely a quantitative science and it is with the quantitative aspects of chemistry and biochemistry that the mole concept is concerned.

The mole concept is a consequence of the particulate theory of matter. The word 'mole' comes from the Latin word *moles,* meaning a pile or heap. Chemists have come to use the word to mean the amount (pile or heap) of a substance containing a standard number of specified particles. These may be atoms, molecules, or ions. There are, thus, the same number of atoms in one mole of carbon atoms (C) as there are molecules in one mole of oxygen (O_2) or in one mole of glucose ($C_6H_{12}O_6$) or as there are ions in one mole of sodium ions (Na^+) or of chloride ions (Cl^-).

In passing, it may be noted that the mole notation is used in physics as well as in chemistry. Thus a mole of electrons constitutes a Faraday of electricity, whilst a mole of light photons constitutes an Einstein of light.

The number of particles constituting a mole is Avogadro's number. Avogadro (1776-1856) was an Italian physicist who made important contributions to the knowledge of gases–in particular to the relation between the volume of gas and the number of gas molecules present. Avogadro's number is now defined as the number of atoms in exactly 12 g. of the carbon isotope ^{12}C, of which the best estimate to date is 6.02252/± 0.00028 x 10^{23}.

It would have been possible to choose some other large number of chemical particles as a mole, for example 10^{24} or 2^{80}. Chemists chose Avogardro's number because this was originally their estimate of the number of atoms in 16 g. of natural oxygen, the chemical reference unit at that time. In 1961 however, the reference unit was changed to 12 g. of the most abundant carbon isotope which is the isotope designated ^{12}C. The change was instituted because, up till then, physicists and chemists had used slightly different reference scales, both based on the element oxygen and it was felt that a unit based on carbon would avoid confusion between the two scales. For most practical purposes the change makes very little difference, for the number of atoms in 16.0 g. of natural oxygen differs from Avogadro's number by less than one part per million. Avogadro's number is, of course, exceedingly large by everyday standards. Some idea of the number of particles in a mole can be gauged from the estimate that a 'mole' of ordinary glass marbles, spread over the whole of Australia, would form a layer nearly 50 miles thick!

For many practical purposes the mole as a unit is inconveniently large and sub-units such as the millimole (=10^{-3} mole), the micromole (=10^{-6} mole) and occasionally, the nanomole (=10^{-9} mole) are much more suitable.

The recommended symbol for the mole is mol.; recommended symbols for subunits are mmol., μmol. and nmol.

2.2. The mole in terms of grams

The mole is a measure of the amount of a chemical substance computed in terms of the number of unit particles present. By definition, a mole of ^{12}C carbon atoms has a mass of exactly 12 g. A mole of some other element has a smaller or larger mass, depending on the mass of an atom of the element compared with that of an atom of carbon. By means of the mass spectrometer an atom of the magnesium isotope ^{24}Mg is found to have a mass almost twice that of a carbon atom. Consequently, a mole of ^{24}Mg has a mass of approximately $2 \times 12 = 24$ g. Similarly, an atom of the platinum isotope ^{192}Pt has a mass of almost 16 times that of a carbon atom and, consequently, a mole of ^{192}Pt has a mass of $16 \times 12 = 192$ g. On the other hand an atom of the helium isotope ^{4}He has a mass only one-third of a carbon atom so that a mole of ^{4}He atoms has a mass of approximately $12 \times 1/3 = 4$ g.

In general, if the masses of atoms are expressed on a scale which gives the ^{12}C carbon atom a value of exactly 12, it is a simple matter to work out the mass of a mole of any atom. The international atomic weight scale is such a scale. Those who have used the term gram-atom will see from this that a gram-atom (or gram-atomic weight) of an element is the same as a mole of atoms of the element.

In dealing with substances whose unit particles consist of two or more atoms, the combined mass of the component atoms is taken into account in computing the mass of a mole of the substance. An atom of hydrogen has one-twelfth the mass of an atom of carbon. Hydrogen molecules, however, each contain two hydrogen atoms so that a mole of hydrogen molecules has a mass two-twelfths of a mole of carbon atoms, i.e. 2 g.

Obviously, in using the mole notation, care must be taken to define exactly the unit particle to be considered. A mole of oxygen *atoms* has a mass of 16 g. but a mole of oxygen *molecules* has a mass of 32 g. Similarly, a mole of sulphur atoms has a mass of 32 g. but a mole of sulphur molecules (each containing 8 sulphur atoms) has a mass of 256 g.

The mole concept applies equally well in the case of molecules containing different types of atoms. If the chemical composition of the unit is known (in terms of the numbers of each type of atom) the mass of a molecule is obtained by addition of atomic masses. For example, the molecular weight of water (H_2O) is 18 so that a mole of water has a mass of 18 g. Similarly, a mole of bilirubin ($C_{33}H_{36}O_6N_4$) has a mass of 585 g. Those who have used the term gram-molecule (or gram-molecular weight) will see that it is the same as a mole.

More recently, the term gram-formula (or gram-formula weight) has been used in dealing with substances such as crystalline solids in which the atoms tend to form infinite arrays. The formula weight is a term given to the mass of the simplest unit of the array. Thus the formula of solid sodium chloride is written NaCl, though sodium and chloride ions actually exist in a lattice and the formula weight of solid sodium chloride on the atomic weight scale is 23 + 35.5 = 58.5. A mole of sodium chloride is generally taken to mean a mole of sodium chloride 'units' and is thus the same as a gram-formula weight of sodium chloride.

In some instances the exact chemical composition of the molecule is not known. This is especially true in the case of macro-molecular substances such as proteins. The weights of such molecules cannot be computed by adding atomic masses, but must be obtained by one or more physical methods. This approach may be necessary even when the chemical composition is known, if there is aggregation of molecules into larger units.

Sometimes it is necessary to calculate how many moles or millimoles there are in a particular mass of a substance. This is simply the mass in grams divided by the molecular weight.

Thus:

$$\text{number of moles } (n) = \frac{\text{mass (grams)}}{\text{molecular weight}}$$

Urea ($C_2H_4N_2O$) for example, has a molecular weight of 60, so that 2 g. of urea constitute:

$$\frac{2}{60}\text{ mol.} = 0.033 \text{ mol.} = 33 \text{ mmol.}$$

Acetone (C_3H_6O) has a molecular weight of 58, so that 10 mg. of acetone constitute:

$$\frac{0.010}{58}\text{ mol.} = 172\ \mu\text{mol.}$$

It is important for the reader to appreciate that there is one salient difference between computing substances in terms of moles and computing substances in terms of mass. In every reaction there is, for all practical purposes, conservation of mass, i.e., the combined masses of the products equals the combined masses of the reactants. Only in certain reactions, however, is the number of moles of products equal to the number of moles of reactants.

Carbon burns in oxygen to form carbon dioxide and in the reaction, one atom of carbon combines with one molecule of oxygen to form one molecule of carbon dioxide. Consequently, one mole of carbon atoms combines with one mole of oxygen molecules to form one mole of carbon dioxide molecules. Thus,

$$\begin{array}{c} C + O_2 = CO_2 \\ 1 \text{ mol.} + 1 \text{ mol.} \rightarrow 1 \text{ mol.} \end{array}$$

In contrast, the mass of carbon dioxide formed equals the combined mass of carbon and oxygen, i.e.,

$$C + O_2 = CO_2$$
$$12 + 32 = 44 \text{ g.}$$

The overall metabolism of glucose to water and carbon dioxide forms another example, viz.:

$$C_6H_{12}O_6 + 6O_2 = 6H_2O + 6CO_2$$
$$180 + 192 = 108 + 264 \text{ g.}$$
$$1 + 6 \rightarrow 6 + 6 \text{ mol.}$$

Again there is conservation of mass but not of moles.

Carbon dioxide can also be produced by the action of an acid on calcium carbonate, as shown by the equation:

$$CaCO_3 + 2HCl = CaCl_2 + CO_2 + H_2O$$
$$1 + 2 \rightarrow 1 + 1 + 1 \text{ mol.}$$

In this reaction, three moles of products are formed from three moles of reactants but if sulphuric acid had been used instead of hydrochloric acid, there would be more moles of products than of reactants, viz.:

$$CaCO_3 + H_2SO_4 = CaSO_4 + CO_2 + H_2O$$
$$1 + 1 \rightarrow 1 + 1 + 1 \text{ mol.}$$

In biological systems, chemical reactions often involve a series of steps with the production of intermediate products which may, in part, be diverted to side reactions. Nevertheless, it is permissible in various circumstances to consider the overall reaction and in doing this the mole concept can be used to express quantitative metabolic relationships as illustrated in the following examples:

(*a*) acetyl CoA → acetoacetate

2 mol. 1 mol.

(*b*) acetyl CoA → cholesterol
18 mol. 1 mol.

(*c*) glycine + succinyl CoA → haem
8 mol. 8 mol. 1 mol.

(*d*) glucose → lactic acid
1 mol. 2 mol.

2.3. The relation between the mole and the equivalent

The concept of chemical equivalence arose towards the middle of the nineteenth century from observations on the relative weights of substances taking part in various reactions. When elements with common chemical properties were compared, it was found that a particular weight of one element was equivalent in various ways to particular weights of other elements. For example, in comparing the reaction of metals with dilute mineral acid, it was observed that 1 g. of hydrogen was 'released' by 23 g. sodium, 20 g. calcium or 33 g. of zinc. It was found also, from the amounts of these metals combining with 8 g. of oxygen, that 23 g. of sodium, 20 g. of calcium or 33 g. of zinc were equivalent. In similar fashion, the weights of ionic substances combining to form salts were found to be related.

The equivalent weight of an element is defined as the amount (in grams) equivalent to (i.e. combining with or releasing) 8 g. of oxygen or 1 g. of hydrogen. For an ionic species, the equivalent weight is the mass with a total charge equivalent to the charge of 1 g. of hydrogen ions.

The equivalent weight of an ionic species is calculated by dividing the ionic weight by the number of unit charges per ion.

$$\text{equivalent weight} = \frac{\text{ionic weight}}{\text{unit charge per ion}}$$

For example, the ionic weight of the calcium ion (Ca^{2+}) is 40 and its charge is 2. An equivalent weight of calcium ion is thus 40/2 = 20 g. In physiology and medicine, the equivalent as a unit proved too large and instead the milliequivalent (= 0.001 equivalent) has been the everyday unit. Thus a milliequivalent of calcium ion is 20 mg.

By strict analogy with uncharged molecules, a mole of an ion species comprises the ionic weight in grams. It follows consequently that the mole of an ion species comprises the equivalent weight multiplied by the charge per ion. Thus:

one mole (of an ionic species) = ionic weight (in grams)
= equivalent weight x charge per ion (in grams)

Sometimes the reader will be faced with calculating the number of moles from the number of equivalents (or milli-equivalents) of an ionic substance. This is simply the number of equivalents divided by charge per ion.

$$\text{number of moles } (n) = \frac{\text{number of equivalents}}{\text{charge per ion}}$$

Thus, a litre of normal plasma contains about 5 mEq. of calcium ions which amounts to 5/2 = 2.5 mmol. of calcium ions.

There are two main reasons why the equivalents notation has been abandoned by chemists and biochemists. First of all, it has a restricted application. It can be readily applied to substances which combine with or displace oxygen and hydrogen, such as the metals or the halogens and to most ionic substances. The term 'equivalent', however, has little meaning when applied to organic compounds such as ethanol, benzene, or glucose. It is true that these substances will react with oxygen, but a whole series of reactions is possible, each requiring a different amount of oxygen, which makes the concept of equivalence impractical.

Secondly, the chemical equivalence of some substances varies according to the circumstances. Under acidic conditions one permanganate ion is equivalent (as an oxidising agent) to two and a half oxygen atoms, but under alkaline conditions a permanganate ion is equivalent to only one and a half oxygen atoms. A molecule of formic acid (HCOOH) can donate one hydrogen ion as an acid but provides two hydrogen atoms as a reducing agent. A molecule of succinic acid provides two hydrogen ions as acid but requires seven oxygen atoms for complete oxidation. The more complex the molecule the more likely is its equivalence to vary according to circumstances.

2.4. The relation between the mole and the volume of a gas

From his studies on the volumes of different gases which react together, Avogadro came to the conclusion that equal volumes of different gases at the same temperature and pressure contain equal numbers of molecules. Since a mole of gas is computed in terms of numbers of gas molecules, it follows that a mole of one gas occupies the same volume as a mole of any other gas, provided it is at the same temperature and pressure.

It is necessary to take the temperature and pressure into account when considering gases, because both influence the volume of a given quantity of gas. The relation between the volume, temperature and pressure of a gas is expressed by the well-known equation: $PV = nRT$, when P is the pressure in atmospheres, V is the volume in litres, T the absolute temperature (in degrees Kelvin) and n the number of moles of gas present. R is a constant, having the value 0.0820 l atm . $°K^{-1}$. mol^{-1} for an ideal gas (see Appendix 1). A gas has not yet been found which behaves exactly according to this equation, though the behaviour of most is not greatly removed from ideal, especially at pressures below one atmosphere.

The molar volume of a gas is defined as the volume occupied by one mole under specified conditions. At a temperature of 273°K (0°C) and a pressure of 760 mm.Hg (1 atmosphere), the molar volume of an ideal gas is 22.4 *l*. Since most gases under physiological conditions behave almost as ideal gases, this value can be used as the molar volume of gases in everyday clinical computations.

It follows that 1 mmol. of gas occupies 22.4 ml. For example, 25 mmol. of bicarbonate ion could give rise to 25 mmol. of carbon dioxide and this, at a temperature of 0°C and a pressure of 760 mm.Hg would have a volume of 25 x 22.4 ml. If the carbon dioxide were at a temperature T°C and at a pressure of P mm.Hg, the volume would be

$$25 \times 22.4 \times \frac{(273 + T)}{273} \times \frac{760}{P} \text{ ml.}$$

If the gas is saturated with water vapour, as is often the case in physiological or laboratory systems, allowance has to be made for this. The total pressure in the system is the sum of the pressure of the gas and the pressure of water vapour in the system. For example, the vapour pressure of water at 38°C is 47 mm.Hg. Thus, in carbon dioxide saturated with water vapour at 37°C and at a total pressure of 765 mm., the part of the pressure due to carbon dioxide is:

$$765 - 47 = 718 \text{ mm.}$$

In V ml. of gas there are thus:

$$V \times 718/765 \text{ ml. of carbon dioxide}$$

and

$$V \times 47/765 \text{ ml. of water vapour}$$

It should be noted that the concept of a single molar volume does not apply to solids or liquids. In these condensed states of matter there is very significant interaction between atoms or molecules, so that the densities of solids and liquids are only very roughly related to the weights of the component atoms or molecules.

3. SOLUTION CONCENTRATIONS IN TERMS OF MOLES

3.1. The mole-per-litre scale (mol./*l.*)

The concentration of a substance in solution is a measure of the relative amounts of solute (the dissolved substance) and solvent. Using the mole notation, the concentration of solutions is usually described in terms of the number of moles per litre of solution. In the case of dilute solutions, it is often more convenient to describe concentrations in terms of sub-units, *e.g.* millimole per litre or micromole per litre. The mole-per-litre scale was formerly termed the molarity scale and a solution which contained a mole of solute in one litre of solution was termed one molar (1 M.). The terms molar and molarity are no longer recommended.

The following examples illustrate descriptions of concentrations in terms of moles:

(*a*) *'5 per cent glucose'*. This contains 5 g. of glucose per 100 ml. The molecular weight of glucose ($C_6H_{12}O_6$) is 180 so that a mole of glucose has a mass of 180 g. A 5 per cent (w/v) glucose solution therefore contains:

$$\frac{5}{180} \times \frac{1{,}000}{100} \text{ mol./}l.$$

$$= 0.275 \text{ mol./}l.$$

$$= 275 \text{ mmol./}l.$$

(*b*) *Urea in plasma*. Normal plasma contains on the average 0.030 g. (30 mg.) of urea per 100 ml. A mole of urea is 60.0 g.

The urea concentration of normal plasma therefore averages:

$$\frac{0.030}{60.0} \times \frac{1,000}{100} \text{ mol.}/l.$$

$$= 0.005 \text{ mol.}/l.$$

$$= 5 \text{ mmol.}/l.$$

(*c*) *Albumin in plasma.* Plasma albumin has a molecular weight of approximately 68,000. A sample of plasma containing 4.0 g. of albumin per 100 ml. has thus a concentration of:

$$\frac{4.0}{68,000} \times \frac{1,000}{100} \text{ mol.}/l.$$

$$= 0.00059 \text{ mol.}/l.$$

$$= 0.59 \text{ mmol.}/l.$$

Since the mole concept applies to ions, the mole-per-litre concentration scale also applies to ionic solutions. A mole of hydrogen chloride dissolved in water yields a solution containing one mole of hydrochloric acid, in the form of one mole of hydrogen ions and one mole of chloride ions. A solution containing one mole of hydrochloric acid in a litre thus contains one mole of hydrogen ions and one mole of chloride ions per litre.

Some examples illustrate ionic concentrations expressed in mole-per-litre.

(*d*) *Physiological saline.* This contains 0.90 g. sodium chloride per 100 ml. of solution. A mole of sodium chloride (NaCl) has a mass of 58.5 g.

Physiological saline therefore contains:

$$\frac{0.9}{58.5} \times \frac{1,000}{100} \text{ mol.}/l.$$

$$= 0.154 \text{ mol.}/l.$$

$$= 154 \text{ mmol.}/l.$$

The solution actually contains sodium ions and chloride ions and, since one mole of sodium chloride gives rise to one mole of sodium ions and one mole of chloride ions, physiological saline is a solution containing:

sodium ions	154 mmol./*l.*
chloride ions	154 mmol./*l.*

(*e*) *3.8 per cent (w/v) sodium citrate dihydrate* (an *in vitro* anticoagulant solution). This contains 3.8 g./100 ml. The formula weight of sodium citrate dihydrate ($Na_3C_6H_5O_7,2H_2O$) is 294. Thus the solution contains:

$$\frac{3.8}{294.1} \times \frac{1,000}{100} \text{ mol.}/l.$$

$$= 0.1292 \text{ mol.}/l.$$

$$= 129 \text{ mmol.}/l.$$

(*f*) *Magnesium in plasma.* Normal plasma contains about 0.0018 g. (1.8 mg.) of magnesium per 100 ml. and a mole of magnesium ions is 24 g. The magnesium concentration in normal plasma is therefore:

$$\frac{0.0018}{24} \times \frac{1,000}{100} \text{ mol.}/l.$$

$$= 0.00075 \text{ mol.}/l.$$

$$= 0.75 \text{ mmol.}/l.$$

3.2. The mole-per-litre scale in volumetric chemistry

Volumetric analysis is quantitative chemical measurement based on the relative volumes of solutions reacting with one another. This is one of the fundamental techniques in analytical chemistry and has many applications in laboratory medicine, particularly in preparation of standard solutions and reagents and in the analysis of body fluids. Until recently, the concentrations of volumetric solutions were expressed in terms of normality; a solution was termed 'normal' when it contained one equivalent of the substance in a litre of solution. However, with the acceptance of the mole concept, the concentrations of volumetric solutions are now expressed in moles.

The main reason for using the mole-per-litre scale in describing volumetric solutions rather than the normality scale lies in limitations of the chemical equivalence notation (see Section 2.3) on which the normality scale is based. In addition, however, the use of the mole-per-litre scale avoids the confusion arising from different uses of the term 'normal', e.g. by chemists to mean 'containing one chemical equivalent per litre' and by physiologists to mean 'having the same osmotic pressure as body fluids', as in 'normal saline'.

It should be noted that when concentrations are expressed on the mole-per-litre scale, one volume of acid does not necessarily neutralise one volume of base of the same mol./*l.* concentration. It happens that 10 ml. of 1 mol./*l.* hydrochloric acid neutralises exactly 10 ml. of 1 mol./*l.* sodium hydroxide because one mole of hydrochloric acid reacts with exactly one mole of sodium hydroxide according to the reaction:

$$HCl + NaOH \rightarrow NaCl + H_2O$$

$$1 + 1 \rightarrow 1 + 1 \text{ mol.}$$

However, 10 ml. of 1 mol./*l.* sulphuric acid neutralises 20 ml.

of 1 mol./*l.* sodium hydroxide because 1 mole of sulphuric acid reacts with 2 moles of sodium hydroxide.

$$H_2SO_4 + 2NaOH \rightarrow Na_2SO_4 + 2H_2O$$
$$1 \quad + \quad 2 \quad \rightarrow \quad 1 \quad + 2 \text{ mol.}$$

To those who have been trained to think of solution strengths in terms of normality, the failure of one volume of base always to neutralise one volume of acid of the same mole-per-litre concentration may appear to be a practical disadvantage of the mole notation. However, the advantages of the mole notation elsewhere outweigh this, and, in fact, those who perform volumetric analysis find little difficulty in working with this scale.

3.3. The molality (mole-per-kilogram) scale

The mole-per-litre scale expresses concentration in terms of the number of moles of solute in a given volume of solution. Since the volume of a given quantity of the solution depends on the temperature (and to a smaller extent on the pressure) it follows that the mole-per-litre concentration of a solution also depends on its temperature. As a solution is warmed and expands in volume, its mole-per-litre concentration decreases (though at ordinary environmental temperatures, this effect of temperature change is relatively small.)

In the molal scale concentrations are expressed in terms of the weight of solvent, which is unaffected by change in temperature. The molality of a solution is the number of moles of solute per kilogram of solvent.

Because of the inconvenience of weighing liquids rather than measuring by volume, the molal scale is not a practical concentration scale. However, as is described in a chapter to follow, the physical properties of a solution, such as its freezing point, are more directly related to its molality than to its mole-per-litre concentration and it is important for the reader to appreciate the relation of one scale to the other. The number of moles of solute per litre in an aqueous solution is usually less than its molality. When some sodium chloride is dissolved in 1 kg. of water, the volume of the resulting solution at any temperature is always more than 1 *l.* (the volume of 1,000 g. of pure water at the temperature of maximum density is 1.0000286 *l.*). Provided, however, the solute concentration is small, so that practically all the molecules in the solution are solvent molecules and provided the solution is at ordinary environmental temperature, at which the volume of 1 kg. of water is practically 1 *l.*, the difference between the molality of an aqueous solution and its mol./*l.* concentration is very small. Thus, 0.1400 mol. of sodium chloride per kg. in water at 20°C corresponds to 0.1396 mol./*l.* In other circumstances, the difference may be appreciably larger. The greater the concentration of the solution and the higher the temperature, the larger the difference will be. Thus, with sucrose solutions containing about 5 g./100 ml. errors of 10 per cent could result from failure to discriminate between the two scales.

The means of converting concentrations from one scale to the other are dealt with in Appendix 3.

3.4. The mole fraction or particle fraction scale

The most fundamental, but unfortunately the most inconvenient, scale of concentration is the *mole fraction* or particle fraction scale. The mole fraction of one substance in a mixture is the ratio of its number of moles to the total number of moles of all the components. This fraction is then the proportion of the particles present which are of a particular type.

The scale of molality is related to the mole fraction scale since it specifies the number of particles of a solute in admixture with a definite number of particles of solvent (the number in 1,000 g.) though there may be other solutes present.

4. CHEMICAL ACTIVITY IN RELATION TO THE MOLE

4.1. The definition of chemical activity

The chemical activity (*a*) of an atomic, molecular or ionic species in a particular environment may be defined as the tendency of the species to exert its chemical properties. As described in the preceding section, the physical concentration of a chemical species can readily be expressed in terms of the number of moles per litre, or in terms of the number of moles of the species relative to the total number of moles present (the mole fraction). The activity of a chemical species, however, is not always directly proportional to its abundance. Experimental studies with solutions, for example, have shown that the chemical effectiveness of each solute particle is reduced as the concentration of the solution is increased and the solute particles are packed more closely together.

The analogy here is a group of infantry (solute particles). If they are positioned in the field closely together they will constitute a more concentrated force than if they are deployed over a wide area, but their effectiveness per man (activity coefficient) is reduced because they tend to get in each other's way. Moreover, to continue the metaphor, those armed with one type of weapon (representing one species of solute) will not only get in the way of their colleagues similarly armed, but will also reduce the effectiveness of others using different weapons (representing other solute species).

The activity of a chemical species should thus be considered related to, but distinct from its concentration. It may be expressed quantitatively on an activity scale and for each of the physical concentration scales (mole-per-litre, molality, mole fraction) there is a different activity scale. Officially chemical activity scales are without units, but the reader will probably find it helpful to retain the units of the relevant concentration scale. The ratio of the chemical activity to the corresponding physical concentration is known as the activity coefficient (γ). Thus:

Activity (a) = activity coefficient (γ) x concentration.

4.2. Activities of solvents

In the case of solvents, the activity (a) is defined relative to the mole-fraction concentration scale. In an ideal solvent, the activity remains equal to the mole-fraction as the amount of solute increases, i.e. the activity coefficient is constant and equal to 1.0 and this holds, in reality, for most dilute solutions. As solutions become more concentrated, however, the activity coefficient of the solvent tends to become increasingly different from 1.0.

In dilute solutions, the activity of a solvent is directly measured by taking the ratio of the vapour pressure of the solvent above the solution, to that above the pure solvent at the same temperature. This physical measure is found to define exactly the effectiveness of the solvent when it takes part in chemical reactions within the solution. Related properties of solutions, such as the elevation of boiling point, depression of freezing point and osmotic pressure, similarly express the difference between the mole-fraction of the solvent and unity and, consequently, reveal the fraction of the particles present which are not solvent molecules. This is put to use in the measurement of *osmolality,* which will be discussed later.

4.3. Solute activities

In the case of solute, the mole-fraction concentration scale is inconvenient and the activity of a solute is usually defined relative to the molality. However, for aqueous solutions such as body fluids which are most conveniently measured out by volume, the mole-per-litre may be taken as the reference scale. Distinction is made between the two activities scales for solutes by retaining units.

In an ideal solution, the activity coefficient of the solute, is 1.0 but solutions only approach the ideal state at low concentration. At higher concentrations such as those found in body fluids, solutions are not ideal.

At any given concentration, the activity of an ionic solution differs from the ideal to a greater extent than that of a non-ionic solution because of the charges of the particles. The electric field of an ion in solution acts on other ions in its vicinity and a particle distribution develops which is biased in favour of those of opposite charge type and against those of like charge. Thus the ion in question no longer moves independently and its chemical action is modified so that it behaves as if its effective charge is less than the actual value. The chemical activity of a given ionic species is therefore affected by the nature and concentrations of all ionic solutes present in a solution.

4.4. Hydrogen ion activities on the mole-per-litre scale

To illustrate the concept of solute activity, we take as an example hydrogen ion activity. In aqueous solutions a proton (H^+) forms a strong attachment to a water molecule to produce the hydroxonium ion (H_3O^+) commonly known as the hydrogen ion, which is attached in turn to other water molecules. This, however, is not an exclusive property of hydrogen ions for many other simple ions, such as the sodium ion or the potassium ion, form strong associations with water molecules and for most purposes the hydrogen ion can be considered without reference to the associated water molecules.

The Danish scientist, Sorensen (1868-1938), considering the wide range of hydrogen ion activities in aqueous fluids, sought a means of expressing the whole range of hydrogen ion activities on a scale which did not involve negative powers of 10. He conceived the idea of expressing hydrogen ion activities in terms of the logarithm of the mole-per-litre concentration with the sign reversed to make the number positive. This allows most hydrogen ion activities (except those of very concentrated solutions of strong acids) to be expressed on a scale ranging from about 0 to about 14 and constitutes the pH scale. By definition, if aH is the hydrogen ion activity (in mol./*l*.) then

$$\text{pH} = -\log_{10} \text{aH}$$

and, conversely

$$\begin{aligned}\text{aH} &= \text{antilog}_{10}\ (-\text{pH}) \\ &= 10^{-\text{pH}}\end{aligned}$$

The pH notation is widely used to express hydrogen ion activities in biological fluids. Thus, the pH of normal arterial blood lies in the range 7.35 to 7.45, the pH of normal gastric juice usually lies between 1 and 2 and the pH of urine lies between 5 and 8.

Two examples illustrate conversion from one acidity scale to the other.

What is the hydrogen ion activity (aH) of a specimen of blood which has a pH of 7.40?

$$
\begin{aligned}
pH &= 7.40 \\
aH &= 10^{-7.40}\ \text{mol.}/l. \\
&= \text{antilog.}\ _{10}(-7.40)\ \text{mol.}/l. \\
&= \text{antilog.}\ _{10}(-8) \times \text{antilog.}\ _{10}(0.60) \\
&= 4 \times 10^{-8}\ \text{mol.}/l. \\
&= 40 \times 10^{-9}\ \text{mol.}/l. \\
&= 40\ \text{nmol.}/l.
\end{aligned}
$$

What is the pH of a specimen of gastric juice which has a hydrogen ion activity of 20 mmol./*l*.?

$$
\begin{aligned}
aH &= 20\ \text{mmol.}/l. \\
&= 0.020\ \text{mol.}/l. \\
pH &= -\log_{10} 0.020 \\
&= -(\bar{2}.30) \\
&= -(-1.70) \\
&= 1.70.
\end{aligned}
$$

As with other ions, the activity coefficient of hydrogen ions depends on the total concentration of ions present. At low total ionic concentration, the activity coefficient of hydrogen ions approaches unity and the hydrogen ion activity equals the hydrogen ion concentration. At higher total ionic concentration the hydrogen ion activity is appreciably less than the hydrogen ion concentration. For example, in the case of a gastric secretion containing H^+, Na^+, K^+ and Cl^- ions respectively in concentrations of 70, 50, 10 and 130 mmol./*l*., the activity coefficient of hydrogen ions is approximately 0.8. This means that the hydrogen ion concentration of such a specimen, determined by titrating with alkali, is about 20 per cent greater than the hydrogen ion activity (obtained from pH measurement). In plasma, the actual hydrogen ion concentration is likewise appreciably greater than the hydrogen ion activity calculated from the pH, though in this situation the former cannot be readily determined by titration because of the presence of buffers. Hence the need to depend on pH measurement.

4.5. Pressure and the activity of a gas

In the case of a gas, the pressure measures its activity. The theoretical justification of the ideal gas equation:

$$PV = nRT \quad \text{(see Section 2.4)}$$

depends on the forces of attraction between the molecules being insignificant. For gases under the conditions involved in physiological circumstances, this equation may, for practical purposes, be accepted as exact. When gases are highly compressed, however, the average distance between the molecules may be reduced to such an extent that the forces become important and the pressure generated by the molecules striking the walls may be less than the ideally expected value. At high pressures the measured pressure of a gas is only an approximate measure of its chemical activity.

4.6. The importance of chemical activities

In elementary chemistry it is not common for distinction to be made between the activity of a solute and its concentration. Calculations are worked on the assumption that the activity of a solute equals its mole-per-litre concentration. There is, however, always the latent reservation that the solutions must be fairly dilute.

As mentioned above, most body fluids contain sufficient dissolved material to cause a significant difference between the activity and concentration of a solute and it is important that this be recognised. To date, thinking in medicine has been mainly in terms of the physical concentrations of the various ions. However, as was discussed in section 4.4, measurements of pH with glass electrode systems are essentially the measurements of the hydrogen ion activity. From the composition of plasma and from the relation between activity and total ionic concentration, it is possible to calculate the hydrogen ion concentration of plasma from the plasma pH. There would, however, appear to be little advantage in carrying out such a conversion, for there is good reason to believe that the physiological properties of an ion depend on its chemical activity rather than on its concentration and it is therefore more appropriate to assess ionic species in body fluids in terms of activities.

Electrodes are becoming available which respond to calcium, sodium and other ions just as the glass electrode responds to hydrogen ions. These will measure the ionic activities directly and may well give a clearer indication of the ionic balance in body fluids than do the simple concentrations.

The reader should also appreciate that, because the chemical activity of an ion species is dependent on the total solute concentration, it is possible for the activity of an ion species in a physiological system to change while its physical concentration remains constant. Conversely, the concentration of an ion species may change without its activity altering. The importance of such dissociated effects in physiology remains to be determined but they are unlikely to be without significance.

5. OSMOLALITY IN RELATION TO THE MOLE

5.1. Definition of osmolality

The osmolality of a solution is a property resulting from the lowering of solvent activity due to the presence of solute (see Section 4.2). Quantitatively it is defined as the molality of the ideal solution which has the same solvent activity. The significance of osmolality requires general consideration of the colligative properties of solution which are therefore considered next.

5.2. Colligative properties of solutions

When material is dissolved in a solvent, the solvent vapour pressure above the solution is less than that above the pure solvent. Quantitatively, an ideal solution exerts a vapour pressure equal to the mole-fraction of solvent, multiplied by the vapour pressure of the pure solvent, i.e. vapour pressure (solution) = vapour pressure (pure solvent) x mole-fraction.

Ideally the difference between the vapour pressures of the solvent and that of the solution divided by the vapour pressure of the pure solvent, equals the mole-fraction of solute. However, the identity of the dissolved material is not revealed. There may be one solute or many; the solute may be ionic or non-ionic, or a mixture.

The presence of solute manifests itself not only by lowering the vapour pressure, but also in an elevation of boiling point, a depression of freezing point and in osmotic pressure. These phenomena, termed 'colligative properties' by Ostwald

(1860-1910), are directly and quantitatively interrelated. The osmotic pressure relates particularly to the fluid transfer between physiological structures and, not being self-explicit, merits brief description here.

By definition, the osmotic pressure is the pressure which must be applied to the solution at one side of a membrane, permeable to solvent but not to solute molecules and separating solvent from solution, to prevent further solvent from entering the solution. If the solution is confined, the internal pressure increases as the solvent passes from the region where its concentration is higher (pure solvent) to the solution where the concentration of solvent molecules is lower, until the solution reaches osmotic pressure.

The complete understanding of why certain membranes are selective in allowing the passage of only certain molecules, is yet to be reached. Molecular size is certainly a factor and it seems likely that different membranes perform this function by different mechanism. Physiological membranes, necessarily, are generally permeable to dissolved substances of low molecularweight, in addition to water, complicating the formal analysis of the situation somewhat. We would emphasise however that the physiological importance of osmolality is not confined to the question of transfer of water across membranes as a result of osmotic pressure differences. Cellular processes occur in a complex aqueous medium possessing thermodynamic properties critically related to its osmolality. Any factor which alters the osmolality of body fluids tends to interfere with cellular processes.

5.3. Osmolality in relation to molality

The colligative properties are most directly dependent on the ratio of the number of solute particles to the total number of solute and solvent particles present. Thus, the mole-fraction scale would be the most fundamental for expressing osmolality quantitatively. In dilute solutions, however, the number of solvent molecules so greatly exceeds the number of solute molecules that the molality scale (moles of solute per 1,000 g. of water) is generally satisfactory. The mole-per-litre scale is less adequate because the density of the solution may be affected in different ways by different solutes and because the mole-per-litre concentration at the freezing point may differ from that at the temperature at which the sample is taken (see Section 3.3).

In more concentrated solutions, osmolality is no longer quantitatively the same as the molality because of the effect of solute on the activity of the solvent. In these circumstances, osmolality is equated to the molality times the osmolal coefficient, thus

$$\text{osmolality} = \text{osmolal coefficient} \times \text{molality}.$$

If the solution contains more than one solute, the effect should be very close to the sum of the effects due separately to the individual solutes. Since 1 mole of sodium chloride (NaCl) gives rise to 2 moles of ions in solution (1 mole of sodium ions and 1 mole of chloride ions) the osmolality will be close to twice the molality. For Na_3PO_4 (trisodium phosphate) the osmolality will be near to four times the molality.

In physiological systems, two factors must be considered in the calculation of osmolalities from molalities. Firstly, most physiological solutions are not sufficiently dilute for the mole fraction (on which osmolality depends) to be accurately proportional to the molality. Secondly, the solute activities tend to differ numerically from solute concentrations. It was to obviate these difficulties that Gamble (1947) coined the term *osmole*. It is defined as the amount of any mixture of solutes which has quantitatively the same osmotic effect in aqueous solutions as one mole of an ideal non-electrolyte.

5.4. Determination of osmolality

Modern temperature sensitive devices, such as thermistors, have made it easy to determine the depression of freezing points of aqueous solutions below that of pure water. Studies on non-electrolytes, such as sucrose at low concentrations, would lead us to expect a solution containing 1 mole of dissolved substances per kilogram of water to begin freezing at −1.858°C, with proportional values for solutions at other concentrations. However, once again, this is ideal behaviour and few solutions have a depression of freezing point equal to 1.858 times the molality, if the concentration is 0.1 mol./kg. or greater. Rather, the depression of freezing point tends to become steadily less than the ideal as the concentration increases. Osmolality is then defined relative to the ideal:

$$\text{osmolality} = \frac{\text{freezing point depression}}{1.858}$$

Osmolalities of sodium chloride solutions have been carefully determined and the osmolality of an unknown solution is usually obtained by comparing its freezing point with those of standard sodium chloride solutions determined by the same procedure.

5.5. Calculation of osmolality

The osmolal coefficient (see Section 5.3) varies with the nature of the solute, its concentration, with the temperature and with other factors. For ions, it is usually less than unity, especially in more concentrated solutions. For some non-electrolytes, it may be slightly greater than unity. The osmotic effects of many substances have been studied and values for osmolal coefficients are to be found in reference handbooks (see Appendix 5).

The osmolalities of mixed solutions, including electrolyte solutions with two or more ionic species, are estimated by summing the contribution from each solute, i.e.,

(total) osmolality = sum of (molal concentration x osmolal coefficient for each solute species).

Two examples serve to illustrate these relations:

(*a*) A solution of glucose contains 0.30 mol./kg. water and has an osmolal coefficient of 1.012. Its osmolality is therefore:

$$0.30 \times 1.012 = 0.304 \text{ osmol./kg.}$$

and its freezing point is

$$-1.858 \times 0.304 = -0.565^\circ\text{C.}$$

(*b*) A solution of sodium chloride contains 0.15 mol./kg. water. The sodium ions have an osmolal coefficient of 0.94 and chloride ions, an osmolal coefficient of 0.92. The osmolality is therefore:

$$(0.15 \times 0.94) + (0.15 \times 0.92) = 0.279 \text{ osmol./kg.}$$

and the solution begins to freeze at

$$-1.858 \times 0.279 = -0.518^\circ\text{C.}$$

The normal osmolality of human blood plasma is about 288 mosmol./kg. water (see Table IV) and the freezing point is approximately -0.536°C. The osmolality of urine in health is usually in the range 100-900 mosmol./kg. but values outside this range may occur with extremes of body hydration or dehydration.

6. THE MOLE IN EVERYDAY MEDICINE

The final section of this manual deals briefly with two aspects of everyday medicine on which the mole notation has the greatest immediate bearing. The first is the chemistry of body fluids such as plasma, urine or faeces; the second is the composition of the various infusion fluids which are used to correct disorders of electrolyte or acid-base metabolism or in parenteral feeding.

6.1. Composition of body fluids in terms of moles

The concentration of some components of plasma on the mole-per-litre scale is given in Table I. The data are a consensus of the findings of various investigations and are intended only to illustrate how the concentration of various components in plasma can be listed on a uniform scale. In practice, normal ranges often depend critically on the analytical procedure used. Readers, wishing a table of normal ranges on the mol./*l*. scale for everyday use are advised, in the absence of locally issued lists, to construct their own by converting normal ranges expressed in conventional units. The means of doing this from the range and the molecular weight is illustrated in the examples provided earlier in this manual (Section 2). Conversion factors for other components have been listed by Dybkaer and Jorgenson (1967).

In compiling the list given in Table I, no attempt has been made to distinguish between the various forms in which certain components exist. Plasma calcium, for example, exists in a number of different forms such as free calcium ions, calcium bound to albumin and calcium chelated with citrate. The range given in Table I refers to all forms of calcium in plasma. However, the concentrations of discrete fractions of calcium, if they are known, can equally readily be expressed in millimoles per litre. Magnesium in plasma similarly is partly bound to albumin as is bilirubin, which, in addition, is to a small extent present as its diglucuronide.

In the case of macromolecular substances, such as proteins, the exact chemical compositions of the molecules may not be known and conversion factors have to be based on the molecular weight estimated by some physical means. In some instances, it is more appropriate to base the conversion factor on a component of the molecule rather than on the whole molecule. For example, in the case of haemoglobin, the mole-gram conversion factor is based on the iron content. Each of the four haem moieties of haemoglobin contains one iron atom and the conversion factor is based therefore on one-quarter of the molecular weight of the whole molecule, i.e. 64,460/4 = 16,115. It is the number of moles of iron atoms which determines the number of moles of oxygen which can be carried by the haemoglobin and the four haem groups act almost independently in this respect. Similarly, with transferrin, the conversion factor is based on the ability to bind iron. Each transferrin molecule binds two iron atoms but the conversion factor is based on the number of moles of iron atoms (Dybkaer and Jorgensen, 1967).

Enzyme concentrations are usually expressed in units based on the rates of reactions under specified conditions. The International Unit is the number of micromoles of substrate changed per minute and concentrations are expressed in 'units per millilitre' or 'units per litre' according to which gives the more convenient figure. It would be possible to express an enzyme concentration directly on the mol./*l* scale if the molecular weight and physical concentration (g./ml.) were known but the figure might not have much physiological significance because of the effect of activators or inhibitors. Enzyme concentrations have, thus, in general more meaning when expressed as enzyme activities. It should be noted, in passing, that the adoption of International Units does not in itself ensure a universally equivalent scale of enzyme measurements, for the activity of an enzyme in a body fluid such as plasma, even though it is expressed in International Units, usually depends on the nature and concentration of the substrate used and the conditions of the assay which vary from method to method.

TABLE III: NORMAL EXCRETION RATES OF FAECAL COMPONENTS IN TERMS OF MOLES

Some excretion rates are influenced by dietary variation.

Component	*Molecular weight*	*Faecal excretion rate* (mmol./24 hr)
Fat (as stearate)	284	10-20
Nitrogen	14	50-100
Urobilinogen		0.15-0.25

The urine excretion rates of various substances are listed in Table II and data on faecal excretion rates are given in Table III. As in the case of plasma composition, the data listed are intended to illustrate the application of the mole notation and they are not valid as normal ranges in all circumstances.

6.2. Calculated osmolality of body fluids

As was earlier described, the osmolality of a solution depends primarily on the abundance of independent solute particles. Body fluids contain many different substances but, even though the concentration of each is known, the abundance of independent particles of every solute cannot easily be calculated because some particles combine. Thus many components, such as calcium, bilirubin, fatty acid, are partly or largely protein-bound. Fortunately, however, the main components which are responsible for the osmolality of body fluids exist largely as free solute particles which allows an

TABLE IV: CALCULATED OSMOLALITY OF AVERAGE NORMAL PLASMA

Component	*Concentration* (mmol./*l.*)	*Molality*‡ (mmol./kg.)	*Osmolal*§ *coefficient*	*Osmolality* (mosmol./kg.)
Sodium	140	148	0.94	140
Chloride	100	106	0.92	98
Bicarbonate	23	24	0.75	18
'Other electrolytes'	7*	7	0.90†	6
Glucose	5	5	1.01	5
Urea	3	3	0.99	3
'Other non-electrolytes'	15*	16	1.00†	16

* Estimated.
† Assumed value.
‡ From the concentration as described in Appendix 4. The density of plasma taken as 1.035, the total solid content as 0.075 g./ml.
§ Each ion has been assigned a separate value (Yardley, 1958).

estimation of osmolality to be derived from knowledge of the concentration of these components.

The process of estimating the osmolality of plasma is illustrated in Table IV. Concentrations have first to be converted from the mol./*l* to the molar scale as is described in Appendix 4. The molality of each component has then to be multiplied by the osmolal coefficient for that component (see Section 5.3) and finally, the contributions of each component have to be summed. Osmolal coefficients for the important constituents of plasma and urine have been tabulated by

Robinson and Stokes (1959). It must be remembered that these coefficients depend on concentration, so that values appropriate to plasma are not necessarily applicable to urine, and also that some urines contain significant amounts of unknown osmotically active material (Yardley, 1958).

6.3. Infusion therapy in terms of moles

It is fortunate that most of the major ionic components of infusion fluids are monovalent ions so that quantities in milli-equivalents are numerically identical with quantities in millimoles. Of the commonly administered ionic substances only calcium forms divalent ions and, consequently, quantities of calcium ions in milliequivalents have to be divided by two to obtain millimoles (see Section 2.3).

A list of gram-mole conversion factor for various substances used in infusion therapy is given in Appendix 5. It is, of course, possible to compute the dosage of most drugs in terms of moles but details of this are outside the scope of this manual, though it is clearly a logical development in the use of moles.

As we pointed out in the introduction, it is unlikely that the description of plasma, urine and faecal chemistries in terms of moles will lead to an immediate improvement in the standard of patient care. The initial reaction of many practising physicians and surgeons faced with learning the new scale may well be one of dismay. For some indeed, it will be the second change, for it is not so very long ago that the milliequivalent scale replaced the mg./100 ml. concentration scale in clinical chemistry. Nevertheless, the change to the mole notation in everyday medicine will undoubtedly occur within the next few years. It is our hope that this manual may help with the transition.

Appendix 1: GLOSSARY OF TERMS AND DEFINITIONS

Acid

A substance which, in the presence of water (or similar solvent) releases hydrogen ions (q.v.) being converted, in the process, to a base (q.v.), e.g.

acid		*base*	
HCl	$\rightleftharpoons$	Cl^-	$+H^+$
lactic acid	$\rightleftharpoons$	lactate ion	$+H^+$
H_2CO_3	$\rightleftharpoons$	HCO_3	$+H^+$
NH_4^+	$\rightleftharpoons$	NH_3	$+H^+$

The extent to which hydrogen ions are released varies from acid to acid, being virtually complete for HCl, HNO_3 and $HClO_4$ up to concentrations of 1 mol./*l.*

Activity Coefficient (γ)

As the concentration of a solution is increased, the chemical activity of a given amount of solute diminishes so that, judged on its behaviour in dilute solutions, its concentration appears to be less than it actually is. The activity coefficient (γ) is the ratio of the chemical activity to the actual concentration. The activity coefficient tends to unity at low total solute concentration (see Section 4.1).

Alkali

A substance giving rise on solution in water to an increase in hydroxyl ion concentration, e.g. NaOH, Na_2CO_3, NH_3.

Atmosphere

A unit of pressure: the pressure exerted by a vertical column of mercury 760 mm. long. This corresponds to a pressure of 1.013×10^5 Newtons per square metre (or alternatively, to a pressure of 1.013 bar). A Newton is defined as the force which would cause a mass of 1 kg. to accelerate at 1 m./sec./sec.

Atomic Mass
Often loosely called the atomic weight; the weighted mean, according to the natural abundance, of the isotopic masses (q.v.) of the atoms of an element on a scale which gives a value of 12.0 exactly for ^{12}C carbon atoms. On this scale the atomic mass of natural hydrogen is 1.0080, the atomic mass of natural oxygen is 15.999 and that of natural carbon is 12.011.

Avogadro's Number
This is defined as the number of atoms in exactly 12 g. of the carbon isotope ^{12}C and is the number of unit particles in a mole of a substance (see Section 2.1).

Base
A substance combining with hydrogen ions to form an acid (q.v.).

Density
The ratio of the mass of a body to its volume, usually expressed in grams per millilitre in the case of solids and liquids and in grams per litre in the case of gases.

Equivalent
The weight of substance combining with or displacing 8.0 g. of oxygen or 1.0 g. (more precisely 1.0080) of hydrogen (see Section 2.3). For ions, the weight equivalent to 1 g. of hydrogen ions. The term equivalent as a unit is no longer recommended.

Faraday
A unit of electrical charge; the charge associated with a mole of electrons. Not to be confused with a Farad–the unit of electrical capacitance.

Formula Weight
The sum of the atomic masses of the component atoms in the empirical formula for a compound (see also molecular weight). Formula weight is used in dealing with substances which exist, at least in part, as polymers or aggregates of uncertain molecular formula. Formaldehyde, for example exists as a

mixture of molecular species represented as CH_2O, $(CH_2O)_2$, $(CH_2O)_3$, etc. However, for computational purposes, formaldehyde may generally be considered to have the formula CH_2O. Formula weight is also used in dealing with ionic compounds such as NaCl or Na_2SO_4 which exist in the solid state as continuous patterned ionic arrays, rather than as discrete 'molecular' units.

Gas Constant
The constant factor (R) relating pressure, volume, temperature and the number of moles of a gas through the equation $PV = nRT$. $R = 0.082$ litre atmospheres per mole per degree.

Gram
A subunit (1/1000) of the kilogram (q.v.): symbol g. Subunits of the gram are: mg., μg., ng.

Gram-formula Weight
The amount of a substance having a mass in grams equal to the formula weight; used in defining a mole of polymers, ionic solids and of aggregates and other substances of uncertain or variable molecular complexity. This is synonymous with the mole of the specified group of atoms.

Gram-molecular Weight
The amount of a substance having a mass in grams equal to the molecular weight. This is synonymous with the mole.

Hydrogen-ion
The term commonly given to the solvated hydrated proton; in aqueous solution, the proton is firmly bound to at least one water molecule.

Hydronium or Hydroxonium Ion
A more descriptive term for the water-bound proton or hydrogen ion (q.v.), H_3O^+.

Ideal Gas
A hypothetical gas which would obey the gas laws exactly at all temperatures and pressures. Such a gas would consist of

perfectly elastic molecules of zero volume and having no forces of attraction between them. For practical purposes, oxygen, nitrogen and carbon dioxide behave ideally at pressures below one atmosphere and at normal temperatures. All gases approach ideality at low pressures.

Ideal Non-electrolyte
A hypothetical substance which would dissolve without dissociating into ions (i.e. a non-electrolyte) giving a solution for which the osmolality is equal to the molality at any concentration. The particles of such a solute would exert no forces on one another. All non-electrolytes approach ideal behaviour at low total solute concentrations.

Ion
An atom or discrete group of atoms carrying one or more units of electrical charge. An aqueous solution containing ions conducts electricity relatively well. In the solid state, ionic compounds including salts consist of a lattice of oppositely charged ions.

Isotopes
Isotopes are forms of an element in which the atoms have the number of electrons and the nuclear charge characterisitic of the element but differ in the nuclear masses (q.v.). It is the nuclear charge and the corresponding electron configuration which determine the chemical properties of an element.

Isotopic Mass
The mass of an atom of an isotope of an element on a scale which gives a value of 12 exactly for ^{12}C carbon atoms. According to convention, the prefix 12 indicates that the isotope has a mass approximately 12 times that of the hydrogen atom.

Kilogram
The S.I. unit of mass; by definition the mass of the international kilogram prototype: symbol kg.

Litre
The unit of volume, now defined (Conférence Générale des Poids et Mesures, 1964) as the volume of a cube with edges one-tenth of a metre, i.e. 1000 c.c. exactly. The old litre was defined as the volume of 1 kg. water at 4°C and 1 atmosphere. 1 old litre = 1.000028 (new) litres.

In the International System of units (S.I. units), the unit of volume is the cubic metre. The Commission on Clinical Chemistry of the International Union of Pure and Applied Chemistry and the International Federation for Clinical Chemistry, however, prefer the litre as a fundamental unit of volume (Dybkaer and Jorgensen, 1967). Symbol *l*., subunits ml., μl., nl.

Metre
The S.I. unit of length, now defined in terms of a wave-length of krypton light. Originally intended to be one ten-millionth part of a quadrant of the earth from the north pole to the equator and, as such delineated by a platinum rod kept in Paris: symbol m. Subunits of the metre are: mm., μm., nm.

Milliequivalent
Subunit (1/1000) of equivalent (q.v.): no longer recommended.

Molality
The obsolete term for the *mole-per-litre concentration*; the number of moles of a solute per litre of solution.

Molarity
The concentration of a solution expressed in moles of solute per litre of solution: no longer recommended.

Mole
The unit of chemical quantity; the amount of substance containing the number of atoms, molecules, ions or units specified by a given formula equal to the number of atoms in exactly 12 g. of ^{12}C carbon atoms; the amount of substance with a mass in grams equal to the molecular weight. Symbol mol.; subunits mmol., μmol., nmol.

Mole Fraction
The ratio of the number of moles of a substance in a mixture or solution to the total number of moles present.

Molecular Mass
The mass of a molecule of a substance on the atomic mass scale.

Molecular Weight
Usually synonymous with molecular mass, see also formula weight.

Normality
The concentration of a substance expressed in equivalents per litre of a solution; insufficiently general for modern use and, consequently, not recommended.

Nuclear Mass
The mass of the nucleus of an atom on the atomic mass scale.

Osmolal Coefficient (g)
The ratio of the osmolality of a solute to its molality. This differs from solute to solute and depends on the temperature and the concentration. For non-electrolytes, the osmolal coefficient is approximately 1.0. Electrolytes in plasma, for example, have osmolal coefficients varying according to solute from 0.75-0.95 (see Table IV).

Osmolality
The molality of ideal non-electrolyte which would produce the same effect on the colligative properties of the solvent (osmotic pressure, freezing point depression, lowering of vapour pressure and elevation of boiling point) as occur in the solution in question.

Osmole
The amount of solute producing in any given amount of solvent, the same osmolality (q.v.) as a mole of ideal non-electrolyte. Symbol osmol., subunits mosmol., μosmol.

Proton
An elementary particle with a positive charge and a mass of 1 on the atomic mass scale; often synonymous with hydrogen ion.

Relative Density
The ratio of the density of a substance to the density of a reference substance (usually water) under specified conditions.

S.I. Units
The name International System of Units (S.I. units) has been adopted by the Conférence Générale des Poids et Mesures for the coherent system based on the units: metre (length), kilogram (mass), second (time), ampere (electric current), kelvin (thermodynamic temperature) and candela (luminous intensity). A seventh unit, the mole, has been recommended for inclusion by the International Union of Pure and Applied Chemistry. In this manual the litre has been retained as the unit of volume as recommended by the Commission on Clinical Chemistry and the gram as the unit of mass to avoid the absurdity of mkg.

Solute
The dissolved substance.

Solvent
The material (usually liquid) in which substances are dissolved to form solutions.

Specific Gravity
See relative density.

Appendix 2: SOME ATOMIC WEIGHTS

bromine	79.9
calcium	40.1
carbon	12.0
chlorine	35.5
copper	63.5
fluorine	19.0
hydrogen	1.0
iodine	126.9
iron	55.9
lead	207.2
lithium	6.9
magnesium	24.3
nitrogen	14.0
oxygen	16.0
phosphorus	31.0
potassium	39.1
sodium	23.0
sulphur	32.1
zinc	65.4

Appendix 3: CALCULATION OF MOLALITY FROM MOLE-PER-LITRE CONCENTRATION

1. Molalities of simple solutions.

With simple solutions containing only one solute, the conversion of concentrations from the mol./*l* to the molal scale can be made if the molecular weight (MW) (or formula weight) of the dissolved substance and the density (ρ) of the solution are both known.

If the concentration of the solution (mol./*l*.) is c, it contains c × MW grams of solution per litre.

The weight of solvent in 1 litre of solution is the weight of solution less the weight of solute, i.e., ($1000\rho - c \times MW$) grams. There are thus c moles of solute in ($1000\rho - c \times MW$) grams of solvent and the molality, being the number of moles of solute per kg. of solvent, is:

$$\frac{1000 \times c}{1000\rho - c \times MW}$$

For example, a solution of sucrose containing 0.5 mol./*l*. has a density (at 20°) of 1.064; the molality of the solution is thus:

$$\frac{1000 \times 0.50}{1000 \times 1.064 - 0.5 \times 342} \text{ mol./kg.}$$

$$= \frac{500}{893} = 0.55 \text{ mol./kg. solvent.}$$

2. Molalities of complex solutions such as body fluids.

If there are several solute species present, the conversion requires a knowledge of the total weight of dissolved matter (TS) expressed in grams per millilitre. The weight of solvent in

1 litre of solutions is the weight of solution (1000ρ grams), less the total weight of solid matter present (1000 x TS grams). The molality of a solute species of concentration c mol./l is thus:

$$\frac{1000 \times c}{1000\rho - 1000 \times TS}$$

$$= \frac{c}{\rho - TS} \text{ mol./kg. solvent}$$

Plasma has a sodium (ion) concentration of 140 mmol./l., a density (at 20°) of 1.025 and a solids content of 0.08 g./ml. The molality of sodium (ion) in plasma is therefore:

$$\frac{0.140}{1{,}025 - 0.08}$$

$$= \frac{0.140}{0.945} = 0.148 \text{ mol./kg. water}$$

$$= 148 \text{ mmol./kg. water.}$$

Appendix 4: MOLE/GRAM CONVERSION FACTORS FOR SUBSTANCES USED IN INFUSION THERAPY

(See Section 6.3)

Substance	*Conversion factor* (mmol./g.)
alcohol	21.8
ammonium chloride	18.8
calcium chloride	9.0
calcium gluconate	2.2
fructose	5.55
glucose	5.55
glucose monohydrate	5.05
potassium chloride	13.3
sodium bicarbonate	11.9
sodium chloride	17.1
sodium lactate	9.0
urea	16.6

Appendix 5: FURTHER READING

Dormandy, T. L. (1966). Body pH. *Lancet,* **1**, 755.

Dormandy, T. L. (1967). Osmometry. *Lancet,* **1**, 267.

Dybkaer, R. & Jorgensen, K. (1967). *Quantities and Units in Clinical Chemistry.* Including recommendations of the Commission on Clinical Chemistry of the International Union of Pure and Applied Chemistry and of the International Federation for Clinical Chemistry. Copenhagen: Munksguard.

Gamble, J. L. (1947). *Chemical Anatomy, Physiology and Pathology of Extracellular Fluid.* Harvard.

IUPAC, (1967). Division of Physical Chemistry, Commission on Symbols, Terminology and Units. *Manual of symbols and terminology for physicochemical quantities and units.* Draft.

Kieffer, W. F. (1965). *The Mole Concept in Chemistry.* London: Chapman and Hill.

Robinson, R. A. & Stokes, R. H. (1959). *Electrolyte Solutions.* London: Butterworth.

Royal Society Conference of Editors (1968). *Metrication in Scientific Journals.* London: Royal Society.

Wolf, A. V. (1966). *Aqueous Solution and Body Fluids.* New York: Harper and Row.

Yardley, H. J. (1958). The composition of normal and pathological urine with an estimate of the concentration of unanalysed substances. *Clin. chim. Acta,* **3**, 280.